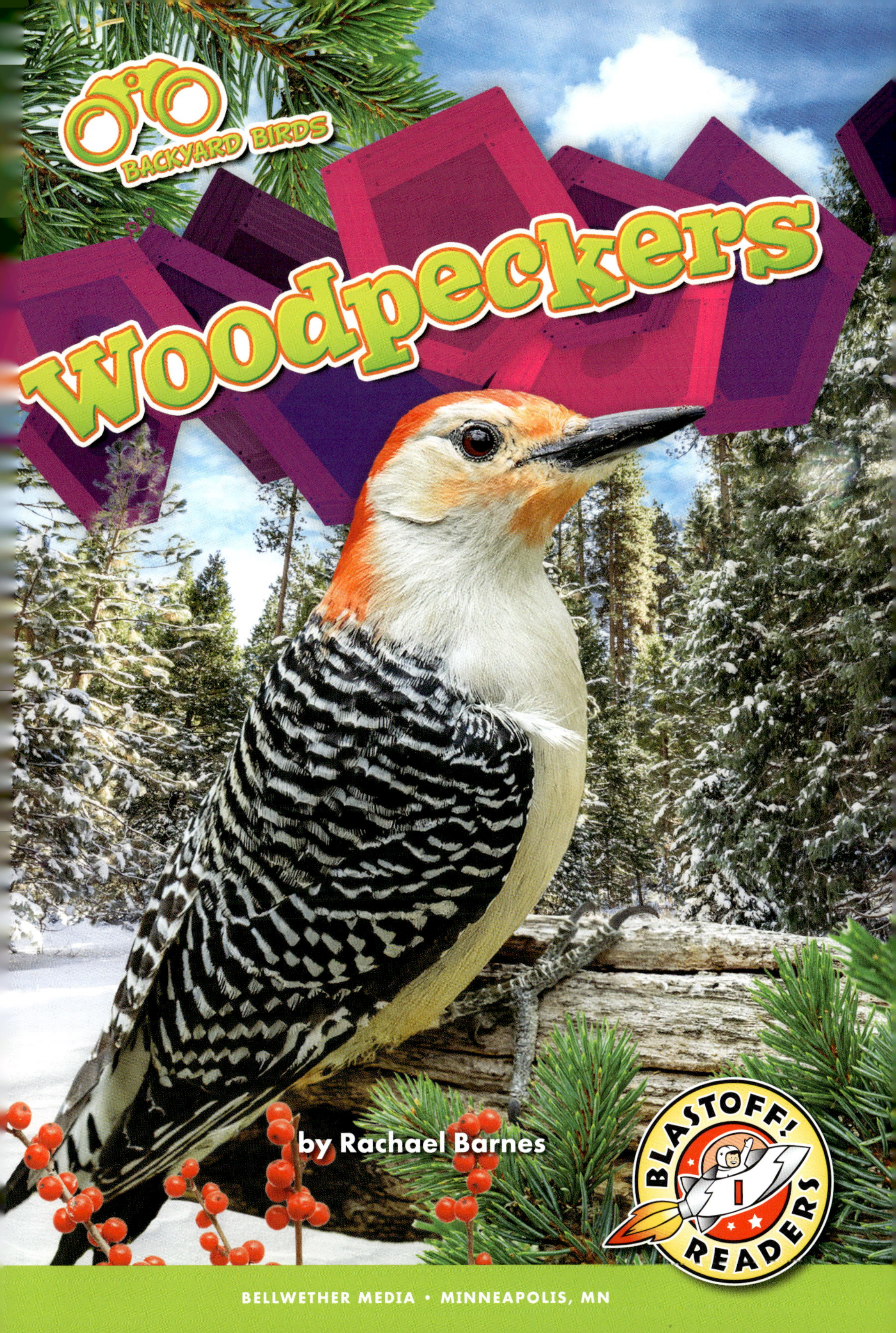
BACKYARD BIRDS
Woodpeckers
by Rachael Barnes
BLASTOFF! 1 READERS
BELLWETHER MEDIA • MINNEAPOLIS, MN

Blastoff! Readers are carefully developed by literacy experts to build reading stamina and move students toward fluency by combining standards-based content with developmentally appropriate text.

Level 1 provides the most support through repetition of high-frequency words, light text, predictable sentence patterns, and strong visual support.

Level 2 offers early readers a bit more challenge through varied sentences, increased text load, and text-supportive special features.

Level 3 advances early-fluent readers toward fluency through increased text load, less reliance on photos, advancing concepts, longer sentences, and more complex special features.

★ **Blastoff! Universe**

Reading Level

Grade K

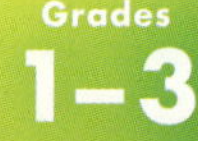

Grades 1–3

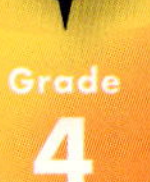

Grade 4

This edition first published in 2023 by Bellwether Media, Inc.

Library of Congress Cataloging-in-Publication Data

Names: Barnes, Rachael, author.
Title: Woodpeckers / Rachael Barnes.
Description: Minneapolis, MN : Bellwether Media, 2023. | Series: Backyard birds | Includes bibliographical references and index. | Audience: Ages 5-8 | Audience: Grades K-1 | Summary: "Developed by literacy experts for students in kindergarten through grade three, this book introduces woodpeckers to young readers through leveled text and related photos"– Provided by publisher.
Identifiers: LCCN 2022002374 (print) | LCCN 2022002375 (ebook) | ISBN 9781644876947 (library binding) | ISBN 9781648347405 (ebook)
Subjects: LCSH: Woodpeckers–Juvenile literature.
Classification: LCC QL696.P56 B37 2023 (print) | LCC QL696.P56 (ebook) | DDC 598.7/2-dc23/eng/20220125
LC record available at https://lccn.loc.gov/2022002374
LC ebook record available at https://lccn.loc.gov/2022002375

Editor: Rebecca Sabelko Designer: Laura Sowers

Printed in the United States of America, North Mankato, MN.

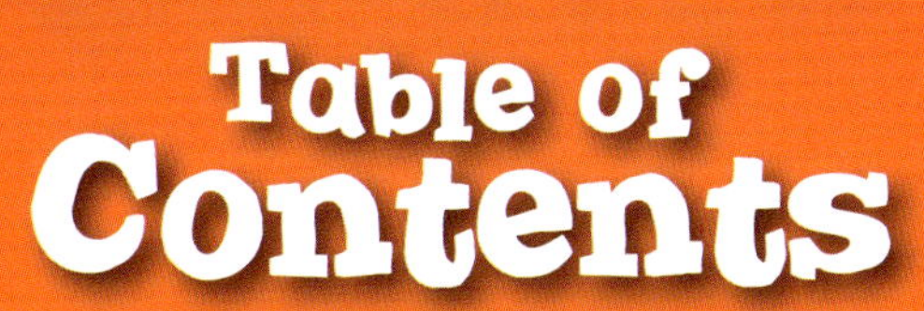
Table of
Contents

What Are Woodpeckers?

Woodpeckers are in the woodpecker family. These birds can be big or small.

All in the Family

northern flicker

pileated woodpecker

yellow-bellied sapsucker

Many woodpeckers are black and white. They often have red feathers on their heads!

Busy Bills

Woodpeckers hang on the sides of trees. Their strong **bills** peck into wood.

bill

These birds make
their nests inside trees.

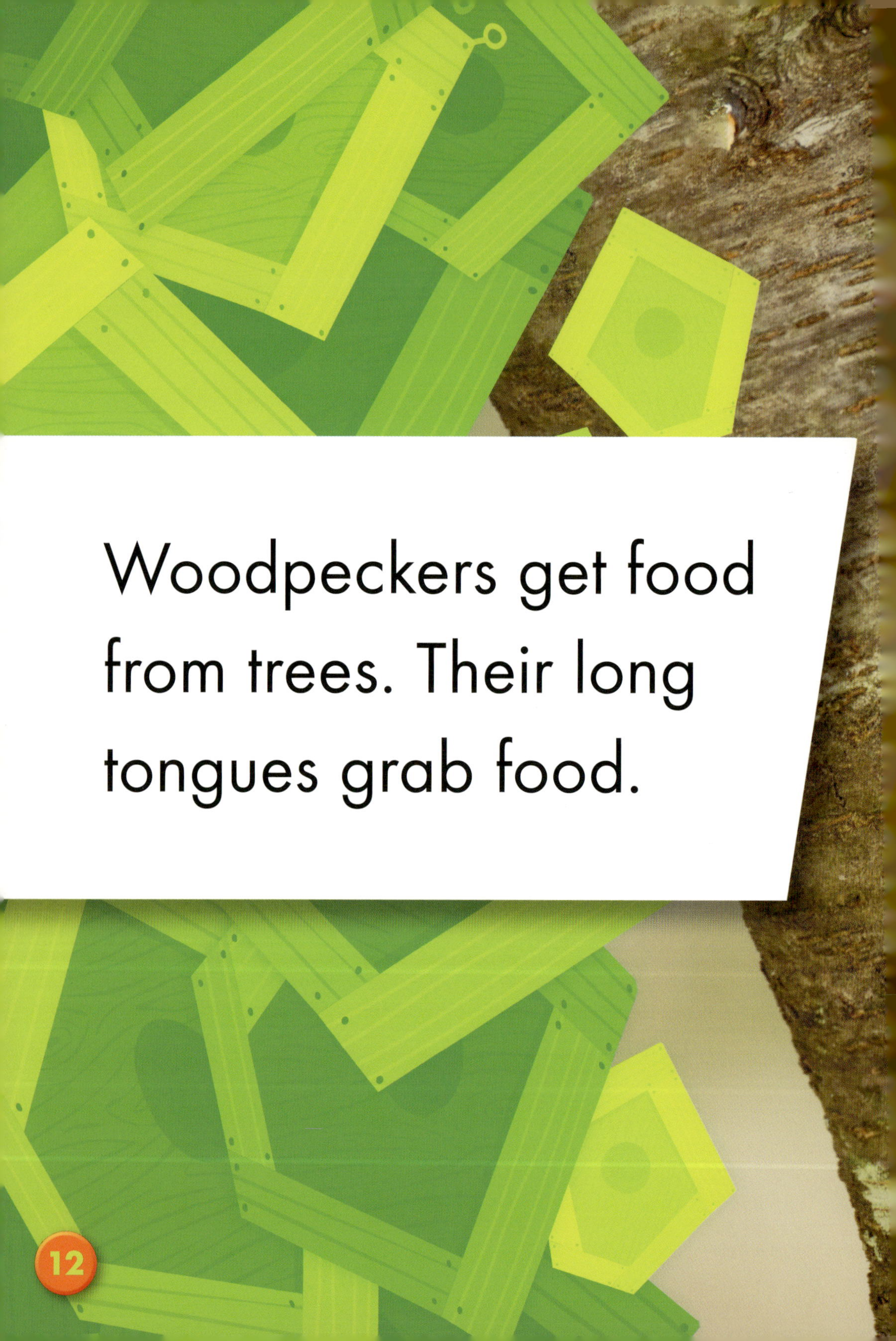

Woodpeckers get food from trees. Their long tongues grab food.

tongue

Woodpeckers mostly eat **insects**.
Some suck up **sap**.
They may eat fruit, too.

Woodpecker Food
insects
sap
fruit

Noisy Neighbors

Most woodpeckers live alone.
Some live in pairs.

Woodpeckers peck to **communicate**. They make loud drumming sounds with their bills.

Woodpecker Sounds
tap
tap
tap

Listen for these birds in forests, deserts, and your backyard!

Glossary

bills

the mouths of birds

insects

small animals with six legs and hard outer bodies

communicate

to send and receive information

sap

a watery juice that moves through a plant and carries food

To Learn More

AT THE LIBRARY

Barnes, Rachael. *Hummingbirds.* Minneapolis, Minn.: Bellwether Media, 2023.

Johnson, Cheryl. *My Backyard Bird Book.* San Diego, Calif.: Puppy Dogs & Ice Cream, 2021.

Statts, Leo. *Woodpeckers.* Minneapolis, Minn.: Abdo Zoom, 2018.

ON THE WEB

FACTSURFER

Factsurfer.com gives you a safe, fun way to find more information.

1. Go to www.factsurfer.com.
2. Enter "woodpeckers" into the search box and click 🔍.
3. Select your book cover to see a list of related content.

Index

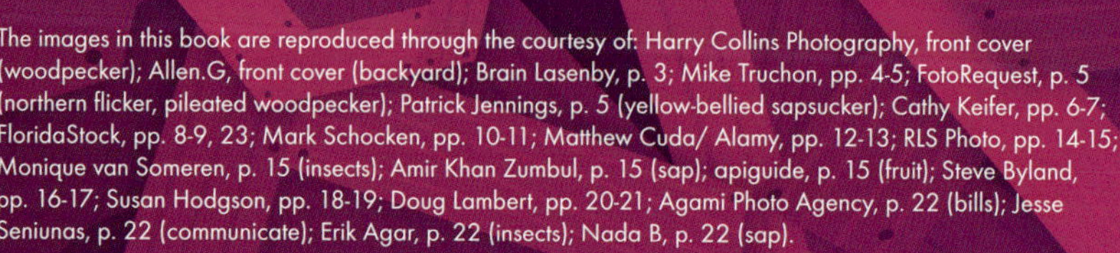

The images in this book are reproduced through the courtesy of: Harry Collins Photography, front cover (woodpecker); Allen.G, front cover (backyard); Brain Lasenby, p. 3; Mike Truchon, pp. 4-5; FotoRequest, p. 5 (northern flicker, pileated woodpecker); Patrick Jennings, p. 5 (yellow-bellied sapsucker); Cathy Keifer, pp. 6-7; FloridaStock, pp. 8-9, 23; Mark Schocken, pp. 10-11; Matthew Cuda/ Alamy, pp. 12-13; RLS Photo, pp. 14-15; Monique van Someren, p. 15 (insects); Amir Khan Zumbul, p. 15 (sap); apiguide, p. 15 (fruit); Steve Byland, pp. 16-17; Susan Hodgson, pp. 18-19; Doug Lambert, pp. 20-21; Agami Photo Agency, p. 22 (bills); Jesse Seniunas, p. 22 (communicate); Erik Agar, p. 22 (insects); Nada B, p. 22 (sap).